Reflections of A Wonderful Journey

THELMA TAPPIN

ISBN 978-1-64349-220-9 (paperback)
ISBN 978-1-64349-221-6 (digital)

Christian Faith Publishing, Inc.
832 Park Avenue
Meadville, PA 16335
www.christianfaithpublishing.com

Printed in the United States of America

I shall attempt to write my biography.

September 12, 2016

I am the middle child of eight children: Alfred, Elaine, Doreen, and Mavis and Thelma, Esme, Roy, and Cedrick, the baby that is said to have taken my mother's life, as she died a few hours after Cedrick's birth. What a thing to lay on a child.

I was born in what I now envision as the heat of the day on a Saturday in mid-August, August 18, 1932. As I grew, there was nothing more enjoyable than having older siblings. I have faint memories of being taken to Sunday school, Bible class, and Ms. Wood's after school programs by my older siblings. Life was so carefree and fun sharing or not sharing. I remember the pranks the older ones played on the younger ones.

My mother was a simple, hardworking, and industrious young woman, with so many children she had to be. Looking back, we never saw her dressed up to go anywhere. The pressing comb was one of the new inventions of the 1920s. I don't think she ever had her hair pressed. I think she had probably resigned herself to her children, and the two eldest girls got it all.

I was about six years old when Elaine and Doreen were being confirmed at St. George's Cathedral. In my young mind, it was quite an event, as when they came home after the service there was homemade ice cream and cake which were quite a treat. I now envision how my mother must have skimped and saved to be able to outfit two girls. All I could think of was the time coming for Mavis and myself to have our turn at confirmation.

I was seven when we were making our first move from the city, Albert Street and South Road, Bourda, to a bigger home in Lodge Village. We felt it was country, with the outhouse in the backyard, the well at the gate, and the La Repentir burial ground one street over, Princess Street. There were no buses in Lodge, only on Vlissengen Road.

With time we were beginning to like 12 Durban Street, Lodge. That was to be short-lived, as my mother died a few hours after the home delivery of Cedrick. What a tragedy. How do you tell a household of children their mother is not coming home from the hospital she was taken to?

Our world stopped, and when it started again, it went into as many different directions as there were children. I don't think there is anything more devastating for a child than the separation from their siblings. It had a profound effect on me. Today I see the effects on some children who are sent to live with estranged parents and stepparents for a better life. Some have tragic endings. I went to live with my godmother Mrs. Irene Payne. I think that was an unwritten code. It was automatic. Nen, as I called her, was a very caring person who even with a family of her own always had it in her heart to reach out to others.

Although I wanted for anything, I went to Sunday school, banquets, and parties with my godsister a few years older than myself. The continuous longing for my siblings made it difficult for me to be happy. I never felt I belonged. My godsister Shirley was a very bright girl and taught me many handcrafts—sewing, knitting, embroidery, and the like. I went to St. Phillip's Anglican School, and although there was no secondary schooling for me, I had a good foundation that paved the way for many future opportunities.

As I left my Nen's home, I went to live with a cousin Mrs. Beryl Elliott. She was a young woman, now I knew how young, with a family. There were two little ones, Joan and Godfrey who was born shortly after I got there. Winston and Maurice were a little older but younger than myself. The little ones did seem to fill that sibling void, but only for a while. Eventually I did end up living with one of my older sisters Doreen. Thank God for the passage of time, as little ones become teenagers and older ones young adults. I was now a teenager. (What's that? I never heard the word.) I was now in my late teens and finally feeling like I had come home where I belong. As I slowly moved into the workforce being responsible for myself, my first obligation to myself was to be confirmed. We were now able on our own to seek out each other and try to reconnect, but the years the locus had stolen from us were gone forever. It would never be the same. Life does not come to a halt for anyone or any circumstance. It keeps right on ticking, and so must we all. Today we all have our own families and have outlived our mother many years over as she died a young woman. We each have our own stories to tell as I am endeavoring to tell mine here.

Life progressed with its many twists and turns of new responsibilities of caring for oneself, thinking of boyfriends and parties, paying attention to politics, world affairs, or just plain economics of everyday life. However, we were rudely awakened by the appearance on the scene of two dynamic, good-looking, bright young men Mr. Cheddi Jagan and Mr. Forbes Burnham who took the masses of British Guyana by storm with their freshly acquired knowledge of politics. This was around 1950. Most of us in my age group, not quite twenty as yet, were not the least bit interested in politics; but when either of these new politicians was having a meeting in our neighborhood, the young ones got a chance to get out at night. Was it to listen to the politicians who were trying to enlighten us or the sweet nothings coming from someone else?

That was a fun time for the young ones.

The adults were taking their politics quite serious. As the election came to fruition, I was slightly underage and could not exercise my franchise. There is a lot more to be said about this period in British Guyana's history but not by me. As I have indicated, I was not politically minded. Life went on with its ups and downs and whatever else in between. In the midst of all of this came the most opportune moment in time for young and not so young women in the country. This started in early 1950s with Jamaica and Barbados. Later Trinidad and Guyana were added. In 1955 through some arrangement with the Government of Ottawa, Canada, and the West Indies and Guyana, an arrangement was made to send women as domestic help to Canada. The first year, young people were not interested in domestic help, let alone living in the home being a domestic help.

That first batch as it became known was made up of mostly mature women closer to the cut of age, which was thirty-five years. Not much attention was given to it, except when a few of the women had such difficulty raising the plane fare of $367.00. It was circled that a couple dropped out on that account and standbys took their place (not confirmed). We did not know this would be ongoing, so when the clarion call came out for new recruits, the younger ones were front and center. Having heard only good reports from those who took the plunge, yours truly and her best friend thought they would give it a try. There was nothing to lose, just making sure one got through. As the selection ordeal started with its meetings, there was fun. Young people would have fun even at a wake. At the meetings, we had different speakers giving us information on where to go, Toronto, Montreal, and Ottawa, and how to behave, as we were ambassadors for our country, the shining light, and could encourage others to follow or stop the flow. One of the speakers was wife of Cheddie Jagan, Janet. She was very interested in the domestics of Guyana as she was one in the government that was very vocal in having the young women take the opportunity. Not all in the government were in for that move. As it turned out, my friend Kathleen got through and I failed along the way. It did not bother me much because of our agreement.

That second batch left, and life with its many changes took place. I was now in new employment at Atkinson Field Guest House. I was away from the city two weeks at a time and was surely able to save some money with my passage in mind. I also enjoyed working at the airport canteen. I would watch with hopeful anticipation as the different planes would taxi in front of the airport, just waiting for my time to board. It did come when the third batch was called and I was present. Many said to me why was I going through that again. However this time, an older sister and the youngest sister were also in the race. I guess they were saying the same thing I had said before—not to worry just as one got through. I was the lucky one, the middle child. This time around, the government was also lending the women the passage which took quite a burden off the shoulders of many. As for me, I now had many sources which I did not have before. Just be patient and wait on God's time. I might have been one not able to raise the fare. My sisters did not have to worry again.

The third batch left Guyana on November 9, 1958. The plane flight was quite an experience. We started feeling the cold somewhere along the way. You should have seen us, huddling under whatever we thought would give us some warmth. Remember these green horns had no idea of what they were going into. Looking back, we were a sore sight on the plane. Luckily, it was just us. I don't remember any others on the plane. I think the reality of where was I going was beginning to set in. We were. At least most of us had never left home to travel anywhere in the country by ourselves. This was autumn, the beautiful fall season.

Goodbye and good luck

Hon. Mrs. Janet Jagan, Minister for Labour, Health and Housing, saying farewell to one of the 30 domestics at the Labour Department, Kingston, Sunday morning before their departure. *1954*

Although I was writing this on October 8, 2016, in the season of autumn, climate change must be responsible for the gorgeous day of sunny and warm weather. This summer of 2016 was recorded as one of the hottest, with very little rain.

We arrived in Montreal and were put up in a hostel for a couple of days, as November 11 is Armistice Day, a national holiday. Those of us for outside of Montreal could not travel. We travelled by train on November 12, 1958, arriving at the employment office in Toronto where we were already assigned our employers.

I was put into a cab and sent to 6 Highland Avenue, Rosedale, which was the W. C. Pitfield's Residence, a small mansion by my outlook. Now I know Mrs. Pitfield was just a couple of years older than myself. There was a baby of three months old, Elizabeth, and a three years old (going on thirty years), Chip, a boy whom I could say was my boss. When I arrived, Chip was supposed to be down for his afternoon nap, but curiosity to see me was more than he could stand. His mother kept telling him he would see me as soon as he got up. I am sure he did not settle in, but when he got up and we met, I could not believe this was the little fellow who was so intent on seeing me. Chip and I got along fine. He was very bright. I think he had an eye on me as much as I had to have on him. He reminded me daily: "Thelma, this is not your home. Your home is in British Guyana." That he knew very well. Once when the parents were away, I being tired of wearing the uniform decided to wear a dress. As soon as I got downstairs and Chip saw me, he said, "Thelma, why are you wearing that costume?" Up to now I am still tickled by the statement. I told you he was bright. He knew I should be in my uniform. I could not let him see I was laughing, but I had to call one of my friends to share the joke. I enjoyed the baby Elizabeth who grew quite attached to me.

Thursdays were our half day off, and all roads led to the YMCA on College Street downtown, followed by Eaton's Annex and all stops in between—Cresceige, Woolworth, Honest Ed, etc. At the YMCA, we would meet young women from other islands and became friends. In evenings we went to our different ways, some to Cecil Street at Mr. Moore's Donavalon Centre, where they would enjoy the evening of music, dancing, games, and just socializing. Many evenings I stopped at The Peoples Church on Bloor Street. Even riding the subway from Union Station to the end of Eglinton Station up and down passed the time.

We looked forward to our time off on Thursdays and one Sunday a month when we would go to Church. Ours was St. Paul's on Bloor Street. And we would visit each other for the rest of the day. Our time in the service was one year. After which we were free to move on in any direction that suited us. We were now landed immigrants. Most of us took the advantage of going to night school classes or Shaw Street commercial classes. For some, the year was an arduous time to fulfill, and they could not wait to get out. For others like myself, it was putting off the inevitable, sticking with the safety and known (though limited) aspects of what to expect when on your own, as I was also awaiting the sponsorship of my now husband to go through.

That took some time; and I spent three years with Chip, Elizabeth, and two additions—John and David. In 1961, I went home for Christmas and got married at that time. When I returned to Canada, I decided to go out on my own. By then, a few were already showing it could be done. My first job was at Eaton's in the warehouse. I did not spend much time there as eager beaver was showing in every way that she could do more. I was soon transferred to the accounts office as a ledger clerk, and as an opening became available, I made it over in the typing pool. Remember I had honed some valuable commercial skills. This was all before the advent of the computer. Your Eaton's account card bill went out by mail. Cecil arrived in summer of 1963. Life had just begun as we needed bigger quarters. That move was quite an experience for both of us as there were no apartments, only flats mainly in the west end. These flats were mainly owned by Greeks, Italians, Jews, and basically Europeans.

They were the only ones that would grudgingly rent to blacks, and I say that because many times when you would answer an ad in the paper and you arrive and they would see a black person, they would say it's already rented. Many can attest to that, even with jobs. There were many restrictions with the rentals, no visitors after such time and no music. I had lived on Roxborough Drive in the east end in one of those mansion-size homes owned by a German man which he split up in rooms and rented to anyone. I am sure his neighbors were not fond of him.

We were now moving to a flat on Christie Street close to Bloor Street. We wasted no time starting our family. In June 1964, a little girl Beverley came into our lives. The baby was now nine months old when we were given notice to move, as a family member was coming in from Italy. As I left home that Sunday morning in tears looking for a place to rent, God provided a flat a few doors down on the other side of Christie Street. And I could not believe I did not walk a half block. We moved, and sure enough another year into our tenancy, we got another notice to move. This time the place was being sold. The baby was now two years old with her brother on the way. We decided to swim or sink with home ownership. It was not by design that we were led back to the east end, 576 Balliol Street with Bayview two blocks east and St. Cuthbert's Anglican Church at the head of Balliol on the east side of Bayview (Leaside). Cecil loved the west end as everything was in walking distance on Bloor Street. We were there for the opening of the east to west subway in 1965 and took the free ride with the baby that Sunday afternoon. Balliol Street was very accommodating as Doreen, my sister, and her family moved in upstairs, and Gordon was born six weeks after the move. It proved to be quite a convenient arrangement.

Now in a new neighborhood not knowing anyone, with two babies, babysitting was quite a challenge. I resigned from my job to be at home. That was not an easy decision or a welcomed one, but once made it was a matter of looking at alternatives to make ends meet. I tell you so much of our upbringing in British Guyana is responsible for our many decisions and outlook on life. We knew about recycling (make ends meet) (cut and contrive) long before the word recycling became popular. The nursery curtains with its motifs of animals and children were quickly transformed to accommodate the first room and board person. That worked out well as it brought in much needed revenue.

I learned to paint, and before you knew it, I was putting a new and bright look in every room. Only a splash of paint could do that, with some help no doubt. You see I was making my stay at home a benefit of the purpose. A worthwhile adventure as I knew a couple of persons who were willing to help me out while helping themselves, by having me babysit for them taking my little ones along. Sometimes I would take one child and sometimes both. We would either be picked up or go by cab. One person lived in Rosedale and the other in Forest Hill. I did what I could to make it worthwhile for them by doing little extras, such as cleaning silver and running the vacuum cleaner while the little ones were down for a nap. The person in Forest Hill had a baby girl the same age of nine months as Gordon just crawling. She was chubby and did not move as fast as Gordon. It worked well for as long as it lasted. I was back home in time to prepare my husband's supper with no one any the wiser of my day's activities. It was a couple of days a week.

As we settled into the community and became members of St. Cuthbert's Anglican Church at the head of Balliol and Bayview, the minister was Canon Wright. He took an interest in us and visited regularly; and before you knew it, he was insisting that I get away with the little ones, as he introduced us to Moorelands Family Camp, also an Anglican undertaking. Not knowing anything about camps, the thought of taking two small kids anywhere for two weeks was not an option. I made every excuse I could think of why we could not go. I got out of it that year, but the next year Canon Wright came back more determined that this would be a good break for me and give me some background into camp life for two weeks with kids. I reluctantly agreed.

As I got us ready for two weeks in August and the day approached, I had already decided if it was not for my tolerance, I would be heading home. We would assemble at a church downtown to be picked up by a school bus. As we got there, the many mothers with their little ones were quite striking, all happy faces, the children as carefree as a lark. I guess it was because knowing Mom was right there. I felt better already. The bus took us to the Beaverton Camp in Beaverton. The arrangement was children six years and older would be in cabins by themselves and the younger would be with their mothers in their cabins. I can't tell you how much we all enjoyed this as the mothers were surely given a break from cooking, laundry, and cleaning while still being able to have an eye on our kids while doing their own activities. It was surely a thoughtful and thankful break for at-home mothers with little ones. Our first year at Beaverton Camp was the last year the camp would be there. The next year the camp moved to Moorelands-Kawagama Camp in Dorset, a much bigger site with much more space and activities. The camping experience quickly became our summer holiday to look forward to. Canon Wright was thanked for his introduction of Moorelands Camp to us and did not have to remind us again. When I started back to work, I still went camping with the kids. I remember a group of mothers were on a trip on a mountain when we heard the announcement of Elvis Presley's death on a little transistor radio I had.

I enjoyed being at home with the kids, and the children enjoyed their aunt and cousins Claudette and Bryan upstairs. There was more TV going on upstairs with the kids. That lasted until Gordon was ready for kindergarten at Maurice Cody. Gordon was very ready as he had had a chance of seeing the inside of the school when we walked Beverley to school. By this time, I had scoured the neighborhood and found the perfect babysitter (Granny), a Trinidadian who was looking after her four grandchildren not living far from us or the school. Granny became not a babysitter to my kids but the grandmother they never had on either side. She loved them as much as they loved her. This also signaled my getting back into the workforce.

I decided to go back and finish the data entry course. I had to quit when I was so rudely interrupted being pregnant with Gordon. I completed the course this time and got my diploma, in time to fill some temporary positions with the federal income tax office. After that, some of us moved on to the Ministry of Health OMSIP (*health card*). Those were the days of those big clunky machines where we punched holes in cards, hence key punch. We progressed with the change of different machines even a real to real, before ending up with what we now know as the computer. The health system also went through a couple of name changes like OMSIP, OHSIP, and finally OHIP.

Give life a chance and deal with the sweat. It always gets better. Remember the only constant is change. The children are growing, and work is work. The bills are being paid.

My friend in Montreal called me with what seemed as a wild notion that we go on a trip. This was a three-city show tour—Amsterdam, Paris, and London. The very sound of it made me giddy. I said yes and was soon packing for the trip to Montreal. I picked her up, and we were on our way. First stop was Amsterdam.

We were like two cats in a strange pantry (never see, come see). We could not take in enough of the strange sights around us.

We visited the Van Gogh Museum with his famous paintings and naturally had to have a canal boat ride. Before we knew it, we were boarding the plane for our next stop—Paris, France. We were totally blown away in the great Paris—"Pinch me to be sure I'm here." In 1973 Paris was still like an impossible dream to us, but then we were now Canadians in Canada much closer. We were just so flabbergasted and awestruck. A glass of wine was easier to have with a meal than a glass of water. The preverbal Eiffel Tower loomed luminously as we explored the sight like a couple of overexcited kids. Paris was breathtaking for all it claims to be. Now for the final city, London, we had a day extra here—in my opinion, unmatched.

There was so much more to see. We both had family and friends there. We each called on someone who was anxious to take us out. The famous Piccadilly Square with the pigeons, Madame Tussauds, London Bridge, and so much more—we were just giddy with excitement. For the first time on the trip, we had rice on one of our menus. You know the Guyanese are rice eaters. We did miss it. We were very comfortable in London. We noticed every city had a square. Toronto did not have one at that time. A few years later, it got Dundas Square. London was a perfect jewel. We enjoyed it immensely. I guess having family and friends to call on made a difference. As our trip came to a close from there, we returned home feeling like travelers back to reality.

It's amazing how a little bit of some things could set you up for wanting more. I was soon introduced to a trip going to the Holy Land and Greece by a co-worker who was a member of Rev. Francis Deeper Life Church in Mississauga. I could not say no. This was even more unimaginable than the first trip. Remember I was a little Third World country poverty-stricken girl who would not have even seen these places in her dreams. [You can never tell the luck of a lousy cat (Guyanese).] Everything about this trip was so awesome, to think you are walking in the footsteps of where Jesus might have trod. The Dome of the Rock is a Muslim place of worship which was built in the seventh century making it one of the oldest structures standing in Jerusalem. Muslim tradition teaches that Mohammed was miraculously transported to heaven from the Rock inside the Dome. Jerusalem is considered to be the third most sacred city in the Muslim world. The crusaders thought the Dome was Solomon's temple and made it into a Christian place of worship. We had to take off our shoes and leave everything outside before entering the Dome. The

Wailing Wall is the only remaining portion of the Jewish temple. We left our written prayers in the crevasses. The Holocaust Museum was quite a shocker. Some of the readings were "The intolerance of the Western world toward Jews made it easy for Hitler and Germany to enforce their agenda."

Shopping in the old city, David Street was lined with shops and markets selling their merchandise. The vendor was as young as six and well versed in bartering. We walked the Via Delarosa, the Stations of the Cross. Some of us were baptized in the Jordan River. We also had a good time in the Dead Sea with the salt mud that is supposed to be so health enriching.

Our tour guide was a Jew. Israel was his name, and he was very knowledgeable about both the history and religious aspects of Israel. As we were travelling on Golda Meir Avenue, we learned the tribe of Benjamin settled there. Moses saw the Promised Land from Mount Nebo but could not enter. Joshua crossed the Dead Sea at Gilboa and defeated the people at Jericho, and the walls came tumbling down. We arrived at Qumran and saw the cave where the Dead Sea Scrolls were found. The Bible just comes alive with the knowledge and sight of so much, from ruins to historic sights.

As enthralled as we were, it was time to move on to the next phase of our trip, the three-island Greek tour. The islands each had their own unique charm. One was like a cat island with countless cats as we got off. They were all well stocked with beautiful merchandise at a price, of course. You'd know the prices did not stop the shoppers from shopping.

More importantly was the good time we all had on the boat cruise. There was another group of Japanese tourists, who wasted no time mingling, taking pictures, and joining in the lustrous singing. I would say mother nature was in her glory with sunshine, fresh air, and hearts that gave praise for all we had seen, had taken part of, had shared, and were leaving behind. It was a glorious day on the water.

We stopped in ancient Corinth. St. Paul spent much time there. The Romans occupied Corinth. Apollo was the only Greek god whose name was not changed. The island of Rhodes was visited by Paul. The Acropolis at the top of Corinth was a corrupt place occupied by Romans. We took pictures of the Berna, a place for the people to speak.

The little we saw of Greece was tantalizing, as we had to leave all too soon for our trip home. I could only pray that much of what I had seen and learned would be indelibly etched on my heart. This was the trip of a lifetime.

Doreen and her family moved on purchasing a home in Scarborough, just a matter of time before we would follow. 228 Crocus soon became our residence. The children were old enough and did not need a full-time babysitter. They were now latch-key kids. As we settled into our home on Crocus, we quickly found what was now our church home, St. Andrew's Anglican Church—a thriving all-white church with a sprinkling of blacks. That was in the 1970s. St. Andrew's has become one of the most multicolored churches in Toronto. At Pentecost, the service had no less than twenty-something different languages in reading the scriptures.

The children were fast becoming teenagers asserting their independence. I was reassured it was no longer necessary for me to take them to our usual Saturday afternoon matinee. They would like to go with friends. As dejected as I felt, I did not brood over that too long. I got the volunteer idea and started looking into it. I started out at a nursing home on Dawes Road. It was going quite well except having to catch a bus after nine o'clock at night. was a bit uncomfortable even then when it was much safer than it is now.

After a while I gave it up. With my antennas out, I heard the new Scarborough Grace Hospital was opening and advertising for volunteers. Three of us from work said we would pursue it. I ended up being the only one to follow through. The first Friday night I showed up, there was Barbara Anderson, a Canadian. There was no one to give us any direction as to what to do. We found ourselves being gofers until we were directed of our duties in the emergency department. Happy am I to say we are both still there in different positions going on our thirty-fifth year. I was truly enjoying my Friday night shift in Grace Emergency Department when it dawned on me that I was approaching a significant milestone of my presence in Canada.

There was so much that had taken place in my life and in the country I had left behind. By now most of my family were in Canada. There's an idea of doing something special for Guyana in the form of giving back for the opportunity I was given. As I always say, some of us came as a vacation and some an adventure. For me it was purely an opportunity to grow and make a difference not only in my life but in the lives of many others, and I do feel I have. *Thank God*. As I pondered the idea, nothing on my own seemed worthy. Hence, I started thinking outside the box. There I included others who came around the same time I did and were interested in, let's say, giving back. Many were not interested having their own ventures. However, as I spread the net wider, I did manage to attract Ethel Applewhite, Enid Nestor, Gloria Sampson Moore, Waveney Limerick (deceased), Doreen Christopher, Hilaries Cameron, and Doreen I fill (deceased). As we got together and tried to come up with something that would benefit most of the masses in Guyana, dealing with the Georgetown Hospital seemed the most relevant. I tell you this because I firmly believe that when God directs your motives and movements, you don't have to know it all. Things just fall into place, and people with what you need will show up and are more generous than you can imagine. As we got our message out, we were overwhelmed with prospects of where and whom to contact. My boss at that time was making room for a new stock of modern filing cabinets. As I approached him for those obsolete cabinets, he said he could not give them away, but told me where I could go to get them—*the government warehouse of obsolescence items.* Mrs. Cameron and I made our way down there. We were like kids let loose in a candy store. It was only the reminder of having to have transportation for whatever we choose restrained us. Hilaries knowing the hospital and its needs took off for the kitchen department, where she collected large aluminum pots and basins and many other kitchen utensils.

We collected three IBM typewriters which we had upgraded and outfitted with new ribbons. We got wheelchairs, crutches, canes, and orthopedic pieces. Let's not forget the hospital beds. Talk about big eyes. We wanted it all but had to remember storage space was also needed, and we could leave some for another time. At the same time, Ethel was negotiating with Baxter Pharmaceutical which was generous in giving us three skids of medical supplies. We were now on our way to the biggest giving back story. With the momentum revved up, there was no stopping now. Hence, the Guyanese Pioneer Fund Raising Group was born in May 1988.

The packing of the first shipment, a forty-foot container, was an event. The shipping agent was Mr. Desmond Debarras. We had all hands on board and some. There was bake and saltfish, coffee, and lots of laughs to lighten the packing effort. Remember we were dealing with hospital beds and lots of heavy stuff. I'll tell you raising that seven thousand for shipping was no laughing matter. There again, the Lord always provides. No amount of bake sales or brunches would have done it in that time. As the thirtieth anniversary faded, thirty-fifth was approaching. I still feel like I owe it to myself to do something for each special milestone. Let's say we got the pioneer group for the thirtieth. The thirty-fifth and onward, I was on my own. For this, a bursary was donated to the Church of St. Andrew for a youth moving on to college or university.

In the fortieth was a stained glass window *Blessed Are Ye* in St. Andrew's Church with a plaque that read *To the glory of God in thanksgiving for my adapted country of Canada 1958–1998 and the Caribbean community of the Church of St. Andrew. Given by Thelma Tappin.* In the forty-fifth, there was a dinner for forty-five homeless men, women, and youths. Somehow the youths did not make it. The dinner, regular turkey and ham with all the fixings, was prepared by myself and a few of the pioneer members. It was served by us and our priest in charge Canon Bill Kibblewhite in the parish hall. My fiftieth anniversary in Canada was very special with a dinner dance held at the Arminian banquet hall. As the request was no gifts, donations could be made to any charity the pioneers supported in Guyana and Canada—the Guyana Relief Council, Ruimveldt Children's Aid Centre, Prostate Cancer Canada, and

Sickle Cell Foundation of Canada. In the fifty-fifth was a bursary donation to St. Andrew's Church. I am looking forward to the sixtieth if it be God's will.

In the interim of these milestones, I had the privilege of making a couple of trips worth mentioning. First, the opportunity presented itself for me to go to Africa to visit Iiius, the little boy our family was a sponsor to for several years. Iiius was seven at the time, and his picture to us revealed a little boy's bewildered look. I felt he needed a big hug. I would go and do just that. At this time Father Kadeni, a native of the Sudan, was the assistant curate at St. Andrew's. He was slightly built, very dark, and said to resemble myself. I think he believed that and claimed me as his sister (called me nothing but sister). Through Father Kadeni and his contacts back home, the church women of St. Andrew's sponsored a kindergarten project in his hometown. It now seemed the Lord Himself was making plans for this trip to happen. The assistant curate Father Kadeni was visited by Canon John Kanyika of Nairobi, Kenya, and Rev. Noel Peters from the Sudan. I was introduced to these big guns of the Anglican diocese in Africa as the sister he did not know he had. Right then I got the bright idea of going to Kenya to see Iiius who lived in Dol Dol, Kenya. All I had to do was get in touch with World Vision there, and they would look after my visit with Iiius when I got there. If I got there, why not hop over to the Sudan to check out the St. Andrew-sponsored kindergarten project?

How ambitious. Father Kadeni having compatriots in both Nairobi and Khartoum would be willing to make all arrangements on my behalf for my arrival and accommodation, at the Anglican guest houses in both places. As word got around that I was planning a trip to Africa and would be visiting the St. Andrew's kindergarten project, some said I was brave going alone, and some thought I was stupid. I just asked all to remember me in their prayers. Basically I consider myself a loner. I did not feel intimidated in any way as I felt safe as all arrangements were made from Canada (really). At that time, the fax machine was the instant communicator. There was no email yet. Now Khartoum was like Guyana in those days; and now there was light, no one knew for how long, which meant receiving or sending a fax was not always reliable. It was still better than mail.

As all arrangements were made and I was well prepared with all of my immunization shots, a suitcase of school supplies for Iiius's school, and gifts for him and his family, there came 9-11. What a shocker. We were grounded a couple of days before travel.

As we were given the go ahead, the flight detoured to Detroit instead of directly to Amsterdam. I tell you a lot of praying was at work. The flight from Amsterdam to Nairobi was seven hours and forty minutes. It was very calm, peaceful, and pleasant. KLM treated us well, with much good food to eat. Before we knew it, the pilot was announcing "Get ready for landing." Thank God everything went well. I got through customs without a hitch.

Now to be picked up, there were many signs, but none said Thelma Tappin. I waited and waited. Everyone was leaving with who came to pick them up. I was starting to get concerned but not afraid. A man from the taxi company in the airport saw me waiting and offered me a cab. I said I was waiting to be picked up. After a while he offered help again. This time I asked him to phone Canon John. There was no answer. We phoned the guest house, they had no reservation for me, and the house was full. After some pleading, she said if I wanted to share a room for the night, I could come. I sure didn't mind and went. It turned out my roommate left at 5:00 a.m. She was just making an overnight stop on her way to another state.

Later in the morning, I phoned Canon John's office. Well, was he ever surprised not only was I in

Nairobi, but I was at the guest house. Right away he sent someone to get me. It turned out he did not get the last-minute communique from Father Thomas. However, he was happy I made it out alright. He was going to phone Rev. Noel Peters with all the information of my flight from Nairobi to Khartoum to make sure there was not a repeat, as this time it would not be as easy with the language barrier. I said that it was what I fear. Canon John called, and I was able to speak to Rev. Noel. I left Nairobi at 5:10 p.m. and arrived in Khartoum at 8:10 p.m. and was through customs by 8:30 p.m. This was the only time my suitcase was opened, but not really searched. As I got through customs and made my way out, I tried to position myself so everyone could see me. After some time of waiting and wondering if this was going to be a repeat, I was getting anxious. Everyone was leaving. I positioned myself in front of a store, with bright lights. It's going on to 9:30 p.m. I must do something. As I tried to speak to the security guard, he brushed me off, "No understand." I had to tell myself now, "Don't panic. No tears. Look normal." I did a too good job with that as everyone was asking what part of Kenya I was from. That was before I opened my mouth. However, this young man without English took a letter of invitation I had and tried to help by finding a phone which was another trip, and he told me to leave my luggage there. That took some nerve, but I did and went with him outside the airport to a phone. The answer was one number was out of service and the other was an office number. The next day being Friday, nothing was open. Remember we were in a Muslim country. So I was back to square one. Thank God my luggage was still there. I was still looking for my pickup. It's now 10:00 p.m. I was coming to grips with the fact that I'd be spending the night right there in the airport. I asked and was given a chair. At least I was comfortable. I ate on the plane, the surroundings were bright stores around, and a plane just came in, so people were passing. In the meantime, I was pulling out all the affirmations I knew and repeating them to keep out the negative powers that were ready to overwhelm me.

I said, "Lord, I know You'll send someone to help me. Please give me the patience to wait." Sure a young man passed by and said something. I did not respond. He followed up with "Welcome to Sudan." I casually said "Thank you." He's gone. I suddenly caught myself. He spoke English. He's gone. I'd lost my chance for help. What could I do but resign myself to my faith. At least I was sitting and praying hard. Lo and behold, the young man was passing back. "Control yourself, Thelma, and don't grab him." So I smiled and beckoned him over. He took my letter, read it, and suggested I take a cab and tell the driver to take me to St. Peter and St. Paul. They would put me up for the night and help me to find my way in the morning. Somehow I did not want to do that. I said I would sit there until morning. He did not think that was a good idea. I asked if he would go outside with me and negotiate with the taxi driver. He agreed. I could tell from the tone it was not going well. He came back to me and said he would go with me. *Hoorah.*

We got to St. Peter and St. Paul, and it's now after 11:00 p.m. As the big gate opened, the watchman took my credentials to the sisters. He came back with the verdict, "We can't have a woman sleep here, but we could take you to the cathedral not far away." There I was turned away from the inn.

My angel knew his way around and took me to the cathedral. As the door opened for us, my angel said "Oh, he has a family," as we saw some children sleeping on mattresses on the floor in this big room, hot and steamy. He then said to me, "You'll be fine. You are with your people (indeed)." Then he left. Only then I felt a bit tearful. I was given a chair and told Rev. Sylvester Thomas would be out shortly. After a while, he did appear just out of the shower in his pajama pants and a towel still wiping his head. I thought to myself I was surely with my people. He took the letter and said he knew everyone there. The next thing I knew, his driver was there to take me to the guest house. Hurrah. It was exactly twelve o'clock midnight when we got to the gate as a Big Ben was striking as we went in. Three men were there including Rev. Noel who exclaimed as he saw me, "Thelma! I was looking for a white person. If you are from Canada, you must be white. I was at the airport as the plane arrived. I had a VIP pass which allowed me to go into the customs area and check the passenger list and saw your name, so I knew you were on the flight." But if he was looking for the wrong color, naturally he couldn't find me. I told you how well I fitted in. He must have looked right past me.

At last I was where I was supposed to be. Thank God for his faithfulness. He has promised never to leave us alone, and I never wavered that help was coming.

I was already noticing the extreme heat, but thank God I had a comfortable air-conditioned room with an adjoining bath. I had a good sleep and had to be awakened. Friday being Muslim day, nothing moved, so it was a day of rest.

On Saturday, I got up early and was ready for the road. Rev. Noel came and offered to reconfirm my flight. He took my passport and ticket and promptly left without me. So much for the road. I settled in, did some embroidery I walked with, and had a good sleep. If nothing else, I should be well rested when I got back. However, when Rev. Noel got back, he told me that some folks were coming to see me in the evening. That turned out to be a press conference without the press. They were the executives of the Little St. Andrew's (kindergarten) and others.

They were eager to tell me all of their problems and troubles and to make sure I saw firsthand, so that in my report it would be more graphic. They also wanted to hear of Canada, and I got a chance to put Rev. Noel at ease by explaining he was not wrong in expecting a white person as we knew that was usually the case. However, it's not that your own did not care or feel your pain but because we were also struggling to help those we had left behind in our own country.

On Sunday, we left for church about 10:00 a.m. The drive was a long way out of the city in an area that was so barren. There was not a blade of grass or a tree in sight, with thousands of makeshift homes that housed the displaced people. People pushed off of their land. There was no water or lights there. I went to see what was called the feeding station. That was the most pathetic sight you could see, an expanse of young women with children under six years old just sitting. I was embarrassed to take pictures, but they seemed happy enough and didn't mind. We got back, had something to eat, and got ready for evening service at the cathedral with Rev. Sylvester Thomas (my rescuer). I was glad as it gave me a chance to take some school supplies and goodies for the kids that were sleeping the night I was rescued there. The service was good and well attended, and I had a chance to bring greetings from St. Andrew's.

Monday, the big day, was finally here. We were going to the Little St. Andrew's kindergarten. This was another trip. However as we arrived, the children being prepared no doubt broke out in song *Welcome, Mama Thelma!* in their language with more words added. It was quite touching to see the little faces beaming with love and excitement in their way of welcoming this stranger. We had a few moments of sheer delight as I opened the bags of much school supplies, small toys, candy, cookies, gum, bath soap, face cloths, and Band-Aids. The visit was short, but I was sure they'd remember something of that morning for some time to come. As we were about to leave, I was presented with a bouquet of wild flowers. I think they had them flown in from FTD.

Little St. Andrew's was the most worthwhile project desperately needed in the desert of Khartoum, which was only surviving like everything else around by the grace of God. People had a faith that defied the awesome challenge of their very existence. I was overwhelmed by their circumstances and their undaunting spirit. Tuesday was my last day in Khartoum, and Rev. Noel was doing his best to have me visit every last who was in the Anglican diocese. A tea party was scheduled for 7:00 p.m. It turned out to be quite a gathering, about thirty people. They sang well, prayed, and gave their well wishes. I responded with greetings from St. Andrew's and my own appreciation for their warm and generous hospitality. I was presented with a staff, which I was told denoted responsibility: I laughed, as it looked more like a cane for my old age. Refreshments were served—pop and cakes (where is the tea?). Within hours I was swept away to the airport for my return to Nairobi.

"Hello, Nairobi. First things first." World Vision was picking me up at 8:30 a.m. the next morning. I was pretty excited. After being confronted with a repeat of Rev. Noel's "I was looking for a white person," there I was again, "I was looking for someone of lighter skin." The two guys in the World Vision van saw me wondering around the compound taking pictures, but did not associate me with the person they came to pick up (a nice way of putting it, is this a joke on me or what?).

We left for a very long and arduous journey only I did not know it. However, the scenery was breathtaking, with hills and mountains, valley and flowers, and the earth with its different shades of bright-red, dark-brown, and white gravel. We passed Mount Kenya. As we left the outskirts of the city, we were entering the desert. The terrain became almost impassable at times. The driver was a good navigator to get through some of those crevasses. After five hours of driving, we arrived at Iiius's school where I met the headmaster and teacher. I took some supplies for the school. We were shown the new school rooms that were in construction.

I met Iiius, and he was the shiest little boy. As I hugged him, the little fellow did not even know how to respond. We then left for his home—the journey of my life. We drove some, until the van could go no further. We had to continue on foot up the mountain I thought we would never reach. It's hard to believe this was the distance the children walk for school every day and I was sure without shoes. As we reached, they were all waiting to welcome us—Mom, Dad, the first wife, and all the thirteen children from both families. The two huts were a little ways from each other. Iiius's mother was the younger and very pretty. This was the Maasai tribe. They deal with beads. They are all decked out in beaded necklaces.

Visit to Africa 2001
Thelma, Iiius and family

We went in the hut and had a bottle of pop which we relished, hot as it was. Guess what. As we left the van, we forgot the suitcase with all the goodies. It was too far for anyone to go get it. Our visit was short as we had to get out of the desert before dark. I tell you that walking down the mountain was something else. The picture would give you an idea. I could have used a donkey, but no such luck. I settled for a staff. Remember this was not a teenager doing this trip, although I had no idea it would be so intense. However, the dad, Iiius, and a couple of the boys came back to the van with us. We dressed Iiius right there in his outfit, sandals, and backpack and took his picture. Iiius's caseworker was on the trip with us and the driver. As we got out of the desert, I was put up at a nice hotel in the town, while the guys spent the night somewhere else. They picked me up at nine o'clock in the morning for the remainder of the trip. I arrived at the guest house at four thirty in the afternoon. That was two days of remarkable and memorable adventure. Thank God for prayers and His protection and guidance, and my steadfast trust in Him. I felt somehow safe. On Saturday Hortencia, Canon John's secretary, was given some time off to take me around. How sweet. We went shopping, and we spent a nice day out.

Thelma, Iius, and his case worker

On Sunday she took me to church at the cathedral. It was their harvest. The church was overrun with produce, everything by the commercial bags. The congregation was matching. The theme was

"Building on 100 Years of Agricultural and Industrial Modernization to Alleviate Poverty in the New Century." It was a two-hour service with no communion. By now the highlights were over, and all I wanted was to get home. Monday and Tuesday were quiet, as I readied myself for my trip home. I now look back with great respect, admiration, and gratitude for the privilege I had—Sudan's airport scare and all—in making that trip at my age. *Thank God.*

I had promised myself when I got back home I would sponsor another child. However, before I could get into that, something came in the mail from the Help the Aged Canada organization pleading for sponsors for elderly persons from Haiti. I decided to go with the Aged. As I looked into it and got the information, the old guy's picture sent to me showed him to be very old and surely needing all the help he could get, but the age confirmed he was younger than myself (seemed as a joke). By the time I got around to taking him serious, someone else had sponsored him. I was given another that looked more like an aged grand, older than myself. The program was very worthwhile, and any help going to Haiti was well worth it. At this time the Guyanese Pioneer Group was giving thanks for another successful annual bake sale and preparing for a pickup of three barrels by Laparkan Shipping to be shipped to the Guyana Relief Council.

You might remember 2005 was a year of horrific national disasters globally. The tsunami in Indonesia was exceptionally devastating in that part of the world. To this day many that managed to get out with their lives are still suffering its losses.

It was also the year Guyana experienced its first real flood scene, as torrential rains poured and poured swelling to the breaking point of the age-old dykes and sluices. As in any tragedy, it brought out the best or worst in most of us. As word got out of the disaster in Guyana, true love of brotherhood engulfed us all. Organizations, families, friends, and concerned individuals from around the globe all responded with heartfelt generosity. We heard the wharfs could hardly hold the barrels. Western Union, Laparkan, and money transfers were swamped with money transfers. That was heartwarming. Not unlike any of the other islands, patriots from far and wide had done their very best in giving back. Just look at the various alumni and different organizations that are working tirelessly in their efforts to create some improvement in various sectors of the Guyanese society regardless of who is in government.

I would say some of us are expressly put here to be busy bodies. I'll be here minding my own business when something or someone would appear saying, "Thelma, let's go." And being the weakling that I am, I'm always ready.

It was summer of 2007 when the opportunity to go on an international build with Habitat in the Dominican Republic came along. I was hoping to do this in Haiti, but the timing was not right, so I settled for the Dominican Republic. This was a couple Susie and Dave, from Ohio, USA, taking the group. As I got and shared all information concerning the trip, Susie seemed to have some doubts of my ability to fit in. Judging from my age, she already had a senior on the team. Remember this was not a holiday. We were going to do some hard work. I must have done a good job putting Susie's fears to rest, as I told her for starters I did not look my age. I had done a local build with my church group on Danforth Avenue, Scarborough, and was quite familiar with habitat in Canada.

From there I was a member of the team. The team consisted of nine females, three males, and Susie and David the organizers, a very nice couple newly retired. The youngest was eighteen-year-old Vivian, a Chinese student from Scarborough, and Kevin, a young man from British Columbia, Canada. All others were from Ohio and other parts of the States. There was one housewife, Marjorie, my roommate; a retired professor from Ohio; a couple of school teachers and a couple of university students; and myself, the grandmother of the group. I am happy to say we all got along very well as they just accepted me as the able-bodied grandmother of the group. My age was never so out-front. The young ones just could not get over me.

One midnight, the oldies had the dorm in a state of panic, as a house frog (a little amphibian) fell from the roof on Marjorie in bed. She woke with such a startle, and there we were creating the biggest stir trying to get the little creature out the door. Everyone was awakened running to see what was going on with the seniors.

Our meals were always tasty and generous, and our laundry was done. Once when the laundry came back, there was a stray G-string, which created some excitement among the guys.

As we arrived in Santiago, we were met by workers of the affiliate in the Dominican Republic. It was a great time of "getting to know you" and orientation meeting. The two seniors shared a room. On our first day, Sunday, we all went to church. Take your pick—Baptist or Catholic.

Our workday started with morning devotion done or read by anyone on the team, followed by breakfast and then being bused to the work site from 8:00 a.m. to 4:00 p.m. Monday to Friday, Saturday being half day. On the work site, we met the real workers, masons, electricians, plumbers, etc. This was not a build of carpentry with hammer and nails. This was a build of masonry concrete and making of rebars. Our task was to prepare the foundation for the next home. The biggest challenge was removing concrete blocks from one spot to closer to the building site. That task started with an assembly line, better known as the block line. This was where you prove yourself and your ability to stand up to the rigors of the construction site. I tell you the grandmother of the team did not disappoint. To pass the blocks from hand to hand, it's done with a swing, and after a time it's done with grunts and groans. It became the Williams grunt and the Sharapova grunt. It added much fun to the task.

After a grueling day in 90-degree weather, we were more than ready for the trip home, where our meal was waiting prepared by the nuns and their kitchen staff. The food was simply all familiar, like saltfish, peas and rice, tuna, etc., done differently and very tasty. There were lots of fresh fruit. No doubt the first night there was some sore muscles, but nothing a good hot shower and a good night's sleep couldn't soothe, just in time for the next day's work.

The next day there were just a few more blocks to be removed. The job was moving piles of dirt by bucketloads in clearing the area. It made the block line seem like more fun. We suddenly had some excitement as a brush fire broke out not very far from where we were. At times it looked threatening but then changed direction. It was still burning when we left.

After supper that night, some friends came over to teach us the Merengue and other Dominican dances. It was great fun, and this senior was the dancing mistress on the floor. They were all so mesmerized by my age. I think it was my agility compared to the other seniors there. I worked even and straight with the young ones.

That night the young ones were going out dancing. Imagine me turning that down. After the night's performance, they were disappointed I was not going. I told them I still had eight more days of hard work so I should be conserving my energy.

At the end of the first week's hard work in 90-degree temperatures, there was a party with the workmen of the site and the Dominican Republic building staff. The highlight of the party was the song I had composed, sung to the tune of "Oh My Darling, Clementine."

Singing the build song I composed

The Words (Chorus)

We are troopers, we are troopers
We're so glad that we are here
We will cherish every memory

As we think of DR build.
Bricks and mortar, nails and hammer
Come together for the build
And today we put the roof on
For to shelter those forlorn
It was a real hit.

Wednesday was a half-day workday. After freshening up and lunch, we were taken on a tour to a research center for chocolates. As the young ones said, it was boring as they did not get any chocolates. We then visited the university campus where we interacted with some of the students informing them of habitat and volunteerism. We were given a chance to use their computer lab to send our emails. We then had a choice of going shopping or home. I was already tired and opted for home. The young ones had to go shopping as there was nowhere to buy their pop or cookies. That was their shopping.

Visit to university
Group Dominican Build

Somewhere along the way, three of us came down with a bug. Vomiting and having diarrhea, I spent all of Saturday in bed with clear fluids. Matt and Vivien were also down, but bounced back faster than I did. We did have good medicine that helped. Luckily it's the weekend. I had Sunday to rest up. I did go to church but rested the rest of the day.

As we are winding down, we did work hard in the extreme heat but were well rewarded as we had a day and two nights in the lovely Wyndham Beachfront all-inclusive hotel. We had a choice of going on a trip by speedboat to the national park or being a beach bum for the day. I chose the park which was quite interesting—a watery park of small islands and caves.

Marg and I working the rebar

One island was a bird sanctuary where all kinds of rare birds could be found. The area was also a nesting ground for whales from all over the world. Our last night was marked with a special dinner and our goodbyes, assessments, and comments on our time together and of each other. To mention one of mine, "Thelma, I admire you for not letting age get in the way of helping others. I think the trip would not have been as much fun without you. I love your outspoken attribute and caring personality. I hope you know you are an inspiration to everyone here. Vivien."

Two weeks had flown by, and we were ready to leave, none the worse for wear and much better for all we had learned and having the privilege of sharing.

Now that you have had firsthand information on the rigors and pleasures of an international Habitat Build, I hope you will be encouraged to try something new and different, even at the local level. Good luck.

You would think coming from a poor country you have seen it all, but I tell you in travelling I have seen poverty such as I have never seen before. I had never seen shanty towns in Guyana or people begging or sleeping on the streets. Yes, in my time we had guys like Johnny Walker, Walker the Niger, Birtie Van, and a few others. I think we just saw them as street enter-tainers, for their antics and crazy talks. Then again my time is not yesterday. I'm sure Guyana is as much in tune as with everything else. There must be a new crop of street people.

There came another church group off to Germany—the Oberammergau Passion Play of 2010. The Oberammergau Passion Play is performed every ten years. It sounded like this was my last chance. I had better take it (I might not be around). The play was first performed in 1634, the result of a vow made by the inhabitants of the village Garmisch in Bavaria, Germany—the vow that if God spared them from the effects of the bubonic plague, then sweeping the region they would perform a passion play every ten years.

The play was an authentically local production, performed entirely by amateur actors who had been residents of Oberammergau for at least twenty years and were of good moral standing in the community. The actors began growing their beards and long hair (no wigs allowed) beginning on Ash Wednesday of the previous year. The costumes were made locally by residents, and the production included live animals.

It was a full afternoon/evening event, from 2:30 p.m. to 5:00 p.m. with a three-hour break, which allowed for ample shopping as the theatre area was overrun with gift shops. Wood carving was their specialty, pieces ranging from religious subjects to toys to humoristic portraits. There was much hand embroidery. The second part started at 8:00 p.m. to 10:00 p.m.

The play was spellbinding, emotional, and thought-provoking, leaving one highly spiritually charged. The orchestra filled moments of heart-dropping scenes.

This was a play of life and death, promised in a moment of mortal threat. So began the history of the Oberammergau Passion Play in 1933, in the middle of the thirty-year war, after months of suffering and death from the black plague—the play of the suffering, death, and resurrection of our Lord Jesus Christ. The first play was done at Pentecost in 1634. The stage was erected in the cemetery above the fresh graves of the plague victims.

That was the highlight of the trip. We went on to see Germany.

As we thought of celebrating Canada's 150th anniversary, we looked back with fond memories to the 1967 celebration in Montreal of its centenary. We were fairly new Canadians not quite having ten years' residence. The whole family made the trip down. Montreal was a new experience for some, but

I had some knowledge of places like St. Catherine's and Cote-des-Neiges. Kathy lived in Montreal. Also my first employers were Montrealers and spent their summer holidays in a quaint little village (Mattice) in Quebec. We would travel down by train, which meant we had a night's sleep over on the train, with meals. That was a novelty to me.

I was pleasantly surprised to find a Trinidadian, a Jamaican, and an English all down with their families vacationing. We spent our time off together.

Expo 67 was an international and universal exposition. It was huge. All I could think of was our little LCP fair, first in the Promenade Gardens and then at Thomas Lands. Guyana had a pavilion here with one of our talking parrots. We heard he or she was cursing even and straight with the workmen on the job. He or she had to be removed hastily. The exposition was like nothing we had ever seen of exhibits and new inventions. Habitat 67 is a model community of housing complex that looked more like a jigsaw puzzle. What a marvel of building ingenuity. Looking back, I cannot remember half of what I had seen fifty years ago. I am humbled to think here I am looking forward to going to Ottawa to celebrate another of my adapted country's significant milestones. What a privilege.

Is it surprising that travel within Canada is as expensive as anywhere else? It is also as beautiful as any place else. I always feel that every place has its own unique beauty and should not be compared. Cecil (my husband) is adamant that Guyana is the most beautiful place you can visit. I have had the privilege of seeing a little of Canada's east, west, and north. In the east Cecil and I drove. We saw a bit of Nova Scotia, the Negro community located in the Barrington Municipal District of Shelburne County, and the museum. Confederation Bridge was in construction.

Quite recently Joyce and I took a bus tour that showed us many more treasured spots: Anne of Green Gables, Cabot Trail, the wonder of Magnetic Hill, Confederation Bridge, and Rita's Tea Room (McNeal) where we had a sumptuous lunch in a beautiful home-style setting. As we were leaving, someone said the lovely teapot on the mantle contains Rita's ashes. It meant nothing to me true or false, but Joyce was a bit alarmed. Luckily we were leaving.

Peggy's Cove was delightful. I was always taken back with a light house. I was lucky to visit our own lighthouse in Kingston, Guyana, right to the top and see the big light.

The bus trip out west Alberta was a girls' trip—Ethel, Doreen, Evelyn (deceased), and myself. Believe me these girls' trip was always a blast. If you know Ethel, you'll understand. There again was the beauty and wonder of nature's gifts to us, harnessed, protected, and cared for by mankind—Butchart Gardens, Lake Louise, Whistler Mountain (the skiers' paradise), Banff Mountains, and more mountains. We had an opportunity to take the cable car that takes you up to the mountain top. Now this calls for speed and agility. As the cable car pulled into the platform, you must be ready to jump on. This looked challenging. Ethel was always up for a challenge and the little one right behind. Doreen and Evelyn were going to be too slow for this. Or so we would have liked to believe. We left them and were on our way up the mountain. As we got to the mountain top surveying the sights and having a hot chocolate, who should appear but the other half of the quartet.

Well I didn't think they ever allowed us to get away with that, and we never stopped laughing.

North was Sault Ste. Marie, a girls' trip driving this time. The drivers were Ethel and Thelma and Doreen and Evelyn the baggage handlers. To get to the Agawa Canyon, we had to take the train. This was an incredible journey on the Algoma Central

Railway through the Agawa Canyon. Unfortunately it was a very cloudy day. At the end of the line, the lookout point could not deliver.

We could only rebound and start our journey back to Sault Ste. Marie.

My, my, how time flies. It has certainly taken its toll, as life moves on and we move with it, we sometimes end up in different directions. Being the youngest of the four probably explains my still being on the road.

When you think of Alaska, one mostly sees lots of snow, glaciers, whales, wildlife, and mountains, but that's not all. These days my travelling partner has boiled down to Bev (my daughter). As we prepared for the Alaskan Cruise we spent a few days with my nephew Lester and his family in Seattle as he would take us to the Ship. There are lots of diamonds in Juneau. Although we did not shop, we feasted our eyes (no charge) in the fabulous Norwegian Pearl, of the Norwegian Cruise Line. Bev and I enjoyed our cruise with all its decadent amenities offered.

Now almost sixty years in my adopted country, it affords me the luxury of numerous memorable and treasured events—the good, the bad, and the indifferent. All can account for my growth, understanding, and appreciation of the vicissitudes of life.

I would like to think that I have used every opportunity to its greatest advantage. Don't we all like to think that we have done it all by ourselves when indeed there are always those unseen and unheard of in the background whom we fail to recognize? Mr. Donald Moore who in the early 1950s was an advocate for blacks with his delegation to Ottawa. The Prime Minister Mr. John Diefenbaker of Canada, Prime Ministers Jagan and Burnham team in Guyana, and many others made it possible for this domestic young woman to be in this country.

I am eternally grateful for all I have achieved and been a part of—from raising a family to being gainfully employed in the provincial civil service for twenty-eight years—for the twenty-five and above years of volunteerism in the Scarborough Grace Hospital, and for the Church of St. Andrew that has fed and nourished my spiritual needs for the past forty years. May God continue to bless my church and church family. They have supported me in every venture I undertook.

Through the years, nothing has given me more joy and satisfaction than the presidency of the Guyanese Pioneer Fund Raising Group. As I mentioned before, giving back is always paramount in my mind. The cooperation of my family, Cecil, Beverley, and Gordon, all made it seem so right.

I was also very blessed with a group of like-minded women and men whose dedication and caring support enabled us to work assiduously in delivering much needed help to organizations in Guyana and globally.

As time marches on, we are all now into the twilight phase of our lives, and its effects are very unkind to some of us. We recognize it's time to pass the torch, take a bow, and exit the stage, except there seems to be no one ready or willing to take the torch. May God in His wisdom and knowledge continue to provide for the needy among us and give us the patience to wait on Him, as if this is of His will we know it will be settled in divine order. God bless.

The Final Chapter

On August 18, 2017, I was celebrating my eighty-fifth birthday. I declared I did not want a big party, but I should invite three couples along with Cecil and myself to dinner somewhere special. As I got around deciding, I settled for the Duncan House. As the evening approached, Cecil announced he was not feeling well and could not go. I called Gordon (my son) who did not readily accept but humored me into thinking he could not turn down his mom on such an occasion. I was very happy he had accepted. The evening was graciously spent, with good food, good company, and light conversation. The ambiance made for a very happy dinner time. Gordon met some of the party for the first time. I was pleasantly surprised he did not spend the evening engrossed in his cell phone, but was somewhat reserved with the elderly company.

It was my evening, and I certainly enjoyed every moment. When we left, we took a few pictures.

As if it was ordained, one week after was the Guyanese Pioneer BBQ. On August 26, Gordon called me around 9:00 a.m. to get the address of the venue for the BBQ. I spoke to him briefly, not even saying my usual "I love you." I was already quite busy putting the last-minute items in place. Winnifred and Bev were faithful supporters. About two hours after Gordon called, the phone rang, and I answered it. The person on the line said, "This is Peel Police. We would like to come and speak to you." Without a blink I said, "I am very busy. If you are coming come right away, I'll be leaving here soon." He said, "Yes, we are coming now." Now I think that was cheeky. I guess I did not think police. I left and went to pick up Joan. On my way I did say to myself I had no traffic violations and soon dismissed it. So true to myself, I am not one to worry in advance of a situation and start thinking the worst. I am the calm and collected one. When the two Peel Police officers arrived, I was out, picking up Joan for more help. The officers then gave the news to Cecil, Bev, and Winnie. They got their reaction and waited for me. As Joan and I walked in, they were tall and imposing in my little living room. Still there I didn't have a trace of anxiety. One officer told me to sit down. As I made my way and had a seat, the officer started talking, and everything thereafter was a blur as I heard Gordon had a fatal accident. My world collapsed, and there I was much like the little girl who lost her mother and had her family split in all directions. Well Miss Cool surely wasn't so cool after that news. I was now told the officers left when I fell apart. I guess they did not dare to watch the cheeky one. "I'm very busy. Come right now." I soon pulled myself together and got the word out. I still don't know how the BBQ survived after that. Thanks to all who were thrust into keeping it alive. The Guyanese Pioneer executive and members, we are most grateful for your help and support in an hour of horrific disbelief.

Gordon had a funeral that had brought out many family and friends, some from afar. It was comforting to be supported by those in person and in spirit. It was a day that the Lord had made perfect in every detail. Everything was in divine order. Our grief was surely made somewhat easier to bear with the support and kindness we received from all of you.

Gordon's stepdaughter (Aubrey) proved she was feeling more confident of her English, as she asked to be one of the eulogists. We were all proud of her.

As I struggled with tears, I was not crying because I felt death cheated on me and robbed me of my son (which it had). I was crying and couldn't help myself because of my mortal sinful nature. How else can I express my pain, my grief, the loss of my love, and the empty seat at the Thanksgiving table? Yes, Jesus wept. I can only look back at how far I have come, not on my own, but with God's grace and trust in Him. As I let go and let God's will be done in my life, as I give thanks for the fifty wonderful years we have had Gordon and the many memories our family and his close friends have shared with us, sometimes I cringe at things he has done unknown to me (us).

Right now, we are taking some solace and comfort in being able to send Gordon's belongings of warm clothing to the Lions Club whose clothing drive would be sending the items out north. The summer clothing would be going to the islands under siege from the recent hurricane that affected Antigua and Barbuda along with food stuff and others.

Gordon Tappin

Who was he? As a toddler he was bright and very active. He was named the speaker of the house by a regular visitor of the home. He could not wait to get to school, as we walked his sister to school. He was soon diagnosed a hyperactive kid. He did the Ritalin thing which he hated, having to take the pill at school. He was an outgoing and likable kid. You could not miss him. He would not be lost in a crowd. I was never told I am beautiful until my son in one of his school projects wrote of his mom as being beautiful—"Beauty only a son could see" (smile). He enjoyed all activities of a youngster, hockey, scouts, camp, church server, and youth group. At that time Rev. Victoria Matthews was a student priest at St. Andrew's. That brood of youths were very active in the church. The sleepover on Easter Eve in the church hall was something to look forward to.

The teenage years were soon upon us. We never knew how lucky we were with no worry of smoking, drinking, or drugs. As school progressed, he was more and more disenchanted with school, resulting in him dropping out, soon to return as he realized that diploma did have some merit.

His Aunt Esme got him his first summer job in Ethel Sterling Factory. I don't know what was his position, but she told us he was driving the machinery, and we knew he had no driver's license. Luckily that job did not last too long as he gave it up for an opportunity to visit relatives with his dad in New York. He soon proved work was never a problem for him. He always had one or two jobs lined up. Currently the entrepreneurial spirit was kicking in. His first venture was being a promoter of dances. This was followed by a different idea as the last failed. He was never discouraged. He always had something else in mind. I am remembering the bathtub caddy.

The next venture was a trip to California for work. He was soon bidding us goodbye. He spent a couple years and decided he did not like it enough to stay. He was coming home. I was very happy. He tried to prepare me for his new appearance, and surely as we met him at the airport, he was wearing his hair in dreads. It was not popular or liked at that time, a new name "Marcelo's." We were thankful it was not a legal change. Since we did not give him that name, we vowed not to use it. He used Marcelo's at work, where no one knew him as Gordon. He got back into his line of work which he knew very well and liked and worked his way to a managerial position. He realized he did not like the confinement of an office. He rather being out in the field. He enjoyed going out west to pick up the new trucks. He was not afraid of hard work or long hours just if it paid well. He took many courses in heavy machinery and driving eighteen wheelers.

It was a relief to all of us when he announced he was going to marry Janet. He said to me, "Mom, you are going to like Janet. She is just like you." I did not ask in what way. I liked the sound of that and left it there.

He said he did not want a big wedding. The city hall would do. We said a small morning wedding in our church of immediate family members would do. It turned out very nicely with the reception held at the Bluffs Restaurant followed by the honeymoon suite at Niagara.

Janet left for her home to the Philippines shortly after to secure her permanent papers and returned with her daughter, Aubrey. Gordon was quite enamored with Aubrey long before she got here, through Skype and the latest technology. Gordon made sure she was registered and ready for school as soon as she got here. It was quite surprising when she asked to be one of the eulogists at his funeral. It meant she was quite confident of her English, which has surely improved.

Gordon would be very much missed by his buddies as Anthony said he was the driving force always pushing them to do something out of the ordinary, like white water rafting. Antony said, "I can't swim. Why did I allow him to talk me into going white water rafting? But I had a good time. It was fun."

This was my son, a conscientious hard worker and always on the go. We shall miss you dearly.

RIP beloved.

About the Author

As a young woman leaving what was referred to as a "Third World country" for the progressive North America, I was told by some of what to expect, such as I would smoke, which I knew would be totally out of character for me. Some said how happy they were for me, as I am the kind that will succeed. Personally I did not allow myself any great expectations. I did pray that I would be able to earn my daily bread doing something other than being a domestic help. I was told too often I could do nothing else. I had to prove myself I could do better given the chance.

As I arrived in Canada, I wasted no time seeking every opportunity by enrolling in night school classes, commercial courses, and community activities in "Red Cross" that kept me grounded.

When I was ready to leave the service, as it was referred to, my first job was as Eaton's Department

Store on Queen and Yonge Streets. In the accounts office, I see myself as a disciplined, caring, thoughtful person, always ready to give a helping hand or hand up. Giving back is my greatest motto, and my reliance on God's daily blessings sustains me.

CPSIA information can be obtained
at www.ICGtesting.com
Printed in the USA
LVHW031053231118
597902LV00001B/1